~A BINGO BOOK~

Plants
Bingo Book

COMPLETE BINGO GAME IN A BOOK

Written By Rebecca Stark
Educational Books 'n' Bingo

ISBN 978-0-87386-441-1

Educational Books 'n' Bingo

Printed in the U.S.A.

PLANTS BINGO DIRECTIONS

INCLUDED:

List of Terms

Templates for Additional Terms and Clues

2 Clues per Term

30 Unique Bingo Cards

Markers

1. **Either cut apart the book or make copies of ALL the sheets. You might want to make an extra copy of the clue sheets to use for introduction and review. Keep the sheets in an envelope for easy reuse.**

2. Cut apart the call cards with terms and clues.

3. Pass out one bingo card per student. There are enough for a class of 30.

4. Pass out markers. You may cut apart the markers included in this book or use any other small items of your choice.

5. Decide whether or not you will require the entire card to be filled. Requiring the entire card to be filled provides a better review. However, if you have a short time to fill, you may prefer to have them do the just the border or some other format. Tell the class before you begin what is required.

6. There are 50 terms. Read the list before you begin. If there are any terms that have not been covered in class, you may want to read to the students the term and clues before you begin.

7. There is a blank space in the middle of each card. You can instruct the students to use it as a free space or you can write in answers to cover terms not included. Of course, in this case you would create your own clues. (Templates provided.)

8. Shuffle the cards and place them in a pile. Two or three clues are provided for each term. If you plan to play the game with the same group more than once, you might want to choose a different clue for each game. If not, you may choose to use more than one clue.

9. Be sure to keep the cards you have used for the present game in a separate pile. When a student calls, "Bingo," he or she will have to verify that the correct answers are on his or her card AND that the markers were placed in response to the proper questions. Pull out the cards that are on the student's card keeping them in the order they were used in the game. Read each clue as it was given and ask the student to identify the correct answer from his or her card.

10. If the student has the correct answers on the card AND has shown that they were marked in response to the *correct questions,* then that student is the winner and the game is over. If the student does not have the correct answers on the card OR he or she marked the answers in response to *the wrong questions,* then the game continues until there is a proper winner.

11. If you want to play again, reshuffle the cards and begin again.

Have fun!

TERMS INCLUDED

agriculture

algae

annual

biology

biome(s)

botany

botanist

bud

bulb

bush

cactus (cacti)

carbon dioxide

cell(s)

chlorophyll

coniferous

deciduous

ferns

flower

forest

fruit

garden

graft

grain(s)

herbs

irrigation

leaf

lichen

kingdom

moss(es)

nectar

nut(s)

perennial(s)

petal(s)

photosynthesis

pistil

plants

Plantae

poison ivy

pollen

pollination

producer(s)

roots

seeds

soil

stamen

stem

tree

tropical rainforest(s)

vegetables

vines

Additional Terms

Choose as many terms as you would like and write them in the squares.
Repeat each as desired. Cut out the squares and randomly
distribute them to the class.
Instruct the students to place the square on the center space of their card.

Clues for
Additional Terms

Write two or three clues for each new term.

_____ 1. 2. 3.	_____ 1. 2. 3.
_____ 1. 2. 3.	_____ 1. 2. 3.
_____ 1. 2. 3.	_____ 1. 2. 3.

agriculture 1. The art and science of farming is called ___. 2. ___ includes the cultivation of the soil, the production of crops, and the raising of livestock.	**algae** 1. Although ___ are similar to plants in some ways, they are not classified as plants. They are in the kingdom(s)Protista. 2. Kelp is a type of brown ___.
annual 1. A plant that completes its life cycle in a single year is called an ___. 2. An ___ germinates, flowers and dies in one year.	**biology** 1. The study of living organisms is called ___. 2. There are two main branches of ___: botany and zoology.
biome(s) 1. A ___ is a major community of living organisms. It is characterized by the dominant plant life and the climate. 2. Some land ___ are tundra, grassland, desert and tropical forest.	**botany** 1. ___ is the branch of biology that deals with plant life. 2. A scientist who specializes in this branch of science is called a botanist.
botanist 1. A scientist who specializes in the study of plants is a ___. 2. Complete this analogy: plants : ___ :: animals : zoologist	**bud** 1. A small bulge on a stem or branch with an undeveloped shoot, leaf, or flower is called a ___. 2. An unexpanded flower is called a ___.
bulb 1. A ___ is the resting stage of a plant that is formed underground. 2. The resting stage of a plant such as a lily, an onion,a hyacinth, or a tulip is called a ___.	**bush** 1. A ___ is a low shrub with a lot of branches. 2. A woody plant that has many low stems that branch out low instead of one main stem is called a ___.

cactus (cacti) 1. A ___ is a succulent, spiny plant. 2. Most ___ are native to arid regions.	**carbon dioxide** 1. This gas is used by plants during photosynthesis. 2. Plants use light energy to combine ___ and water to form sugars.
cell(s) 1. The ___ is the basic unit of life. 2. Plant ___ have a wall made of cellulose.	**chlorophyll** 1. This pigment absorbs light and allows photosynthesis to occur. 2. ___ is a the pigment that gives a plant its green color.
coniferous 1. Trees with cones and evergreen leaves are ___. 2. Pines, spruces and firs are examples of ___ trees.	**deciduous** 1. ___ trees shed their foliage at the end of the growing season. 2. Maples and other ___ trees lose their leaves in the fall.
ferns 1. ___ are flowerless plants with roots; stems; and large leaves, called fronds. They do not produce seeds. 2. These plants do not flower. They have no seeds and reproduce by spores.	**flower** 1. The reproductive structure of a plant is called a ___. The male part is called a stamen. The female part is called a pistil. 2. A ___ consists of a pistil and/or a stamen and often includes petals and sepals.
forest 1. A large expanse of dense growth of trees and underbrush is called a ___. 2. An ecosystem that is dominated by trees is called a ___.	**fruit** 1. The edible reproductive body of a seed plant is called the ___, 2. Apples, pears and peaches are examples of ___.

garden 1. A plot of land where plants are cultivated is called a ___. 2. Many people plant flowers or vegetables in a ___.	**graft** 1. A shoot or bud that has been joined to another plant is called a ___. 2. To ___ is to affix the tissues of one plant to the tissues of another.
grain(s) 1. Another term for cereal grass is ___. 2. ___ are grasses that have been cultivated for the edible components of their fruit seeds.	**herbs** 1. Plants valued for their flavor, scent or medicinal value are known as ___. 2. Some ___ used in cooking are are basil, sage, and thyme.
irrigation 1. The watering of land by artificial means is called ___. 2. In dry regions it is necessary to use ___ to in bring enough water to the area to plant crops.	**leaf** 1. It is an outgrowth of a plant stem. 2. This plant part has two main functions: the manufacture of food through photosynthesis and cellular respiration.
lichen 1. This plant-like organism is made up of an alga and a fungus. 2. The alga and fungus that make up a ___ have a symbiotic relationship.	**kingdom(s)** 1. Natural objects are classified into 3 divisions, or ___: the animal ___, the mineral ___ and the plant ___. 2. In biological taxonomy, ___ ranks above phylum.
moss(es) 1. These plants grow in clusters on rocks, on trees and on the ground. They have a velvety appearance. 2. ___ is a small, low-growing, nonvascular plant that grows in most moist habitats.	**nectar** 1. The sweet liquid produced by some flowers is called ___. 2. This sweet liquid attracts insects such as bees and butterflies.

Plants Bingo

nut(s) 1. A hard-shelled seed is called a ___. 2. Cashews, peanuts and pecans are examples.	**perennial(s)** 1. A ___ plant lives for more than two years. Tulips, irises and most other bulb plants are ___. 2. A plant that lasts several years with new growth is a ___ .
petal(s) 1. A ___ is one of the leafy structures of a flower. 2. ___ are often brightly colored and come in many shapes. All of a flower's ___ together are called the corolla.	**photosynthesis** 1. ___ is the process by which green plants use sunlight to convert water and carbon dioxide into carbohydrates. 2. Chlorophyll is necessary for ___ to take place.
pistil 1. The female, ovule-bearing part of the flower is the ___. 2. The ___ has 3 parts: the stigma, the style, and the ovary.	**plants** 1. ___ are members of the kingdom Plantae. 2. Unlike animals, ___ are capable of making their own food. Also unlike animals, they lack locomotion.
Plantae 1. This is scientific name for the plant kingdom. 2. The kingdom ___ contains about 300,000 different species of plants.	**poison ivy** 1. This woody vine produces a skin irritant that causes an itching rash. 2. Contact with this plant often causes a bothersome rash.
pollen 1. ___ is a fine powder that is produced by some plants when they reproduce. 2. The anthers, which are part of the pistil, carry the ___.	**pollination** 1. The transfer of pollen from the anther to the stigma of a flower is called ___. 2. Bees are important agents of ___.

Plants Bingo

producer(s) 1. As primary producers, plants are at the base of any food chain. 2. Organisms in a food chain can be ___, consumers or decomposers.	**roots** 1. A plant's ___ anchor the plant. 2. A plant's ___ absorb water and minerals from the soil and, along with the stem, store food.
seeds 1. The fruit of a plant contains its ___. 2. ___ are dispersed in many ways. Those of dandelions are dispersed by the wind.	**soil** 1. ___ is the thin upper layer of earth in which the roots of plants grow. 2. ___ is made up of weathered rock and decayed plant and animal matter. Plants are grown in it.
stamen 1. The male reproductive organ of a flower is called the ___. 2. This organ of a flower bears pollen. It is made up of a filament and an anther.	**stem** 1. A plant's ___ provides support for the plant and holds it up to the light. 2. A plant's ___ transports the water and minerals from its roots to the leaves.
tree 1. A ___ is a woody perennial plant with a single main stem and branches that form an elevated crown. 2. The stem of a ___ is called a trunk.	**tropical rainforest(s)** 1. ___ have tall trees and are in regions of year-round warmth. They usually get more than 100 inches of rainfall each year. 2. There are 4 layers in a ___: the floor, the understory, the canopy and the emergent layer.
vegetables 1. ___ are plants that are cultivated for their edible parts, such as stem, leaves, bulbs, or roots. 2. Carrots, lettuce and onions are examples of ___.	**vines** 1. ___ are weak-stemmed plants that need support as they grow. 2. ___ get the support they need by climbing, twining, or creeping along a surface.

Plants Bingo

Plants Bingo

flower	annual	graft	tropical rainforest(s)	roots
bulb	forest	stem	kingdom(s)	leaf
pollen	petal(s)		garden	nectar
seeds	algae	ferns	stamen	grain(s)
herbs	tree	cactus (cacti)	agriculture	fruit

Plants Bingo

tropical rainforest(s)	producer(s)	irrigation	perennial(s)	herbs
grain(s)	coniferous	bud	algae	poison ivy
pistil	tree		carbon dioxide	ferns
kingdom(s)	plants	petal(s)	vines	leaf
fruit	stem	cactus (cacti)	bulb	agriculture

Plants Bingo

tropical rainforest(s)	ferns	kingdom(s)	stamen	pollen
tree	annual	biology	forest	moss(es)(es)
algae	stem		Plantae	biome(s)
petal(s)	pistil	herbs	coniferous	irrigation
agriculture	cactus (cacti)	bulb	vines	graft

Plants Bingo

petal(s)	Plantae	graft	cactus (cacti)	herbs
lichen	coniferous	forest	perennial(s)	pollen
garden	bud		roots	stamen
ferns	deciduous	stem	bulb	biology
agriculture	fruit	nut(s)	chlorophyll	nectar

Plants Bingo

fruit	roots	algae	bud	cactus (cacti)
lichen	ferns	biology	petal(s)	cell(s)
producer(s)	nectar		annual	graft
leaf	Plantae	flower	vines	chlorophyll
kingdom(s)	bulb	photosynthesis	carbon dioxide	garden

Plants Bingo

biome(s)	Plantae	irrigation	producer(s)	nectar
stamen	algae	chlorophyll	forest	pollen
perennial(s)	biology		bud	carbon dioxide
bulb	herbs	vines	nut(s)	garden
grain(s)	ferns	flower	photosynthesis	graft

Plants Bingo

flower	Plantae	pollination	cell(s)	kingdom(s)
grain(s)	graft	tree	annual	pollen
irrigation	stamen		carbon dioxide	botany
petal(s)	coniferous	lichen	tropical rainforest(s)	pistil
cactus (cacti)	bulb	vines	nut(s)	biology

Plants Bingo

garden	Plantae	botanist	stamen	botany
lichen	producer(s)	perennial(s)	graft	roots
pollen	poison ivy		nectar	bud
agriculture	petal(s)	tropical rainforest(s)	chlorophyll	coniferous
stem	bulb	nut(s)	algae	grain(s)

Plants Bingo

carbon dioxide	kingdom(s)	tree	pollen	nectar
chlorophyll	producer(s)	garden	algae	graft
moss(es)	flower		annual	botanist
botany	fruit	herbs	cell(s)	pollination
coniferous	vines	biome(s)	tropical rainforest(s)	roots

Plants Bingo

seeds	tropical rainforest(s)	bud	perennial(s)	photosynthesis
nectar	botany	forest	annual	graft
Plantae	poison ivy		stamen	pistil
herbs	leaf	chlorophyll	vines	moss(es)
bush	fruit	irrigation	grain(s)	garden

Plants Bingo

biology	poison ivy	algae	chlorophyll	grain(s)
botanist	moss(es)	cell(s)	carbon dioxide	forest
lichen	producer(s)		irrigation	tree
bush	pollen	vines	bulb	tropical rainforest(s)
biome(s)	cactus (cacti)	flower	nut(s)	kingdom(s)

Plants Bingo

kingdom(s)	coniferous	moss(es)	stamen	carbon dioxide
tree	stem	producer(s)	nut(s)	lichen
flower	pollination		nectar	perennial(s)
cactus (cacti)	roots	graft	tropical rainforest(s)	annual
poison ivy	botanist	Plantae	biology	botany

Plants Bingo

bush	roots	biome(s)	moss(es)	nectar
producer(s)	botanist	Plantae	carbon dioxide	pistil
stamen	bud		tree	pollination
garden	vines	botany	poison ivy	tropical rainforest(s)
bulb	leaf	nut(s)	flower	cell(s)

Plants Bingo

cactus (cacti)	producer(s)	algae	carbon dioxide	bush
botany	flower	moss(es)	annual	pistil
chlorophyll	stamen		irrigation	bud
leaf	vines	Plantae	biology	biome(s)
bulb	perennial(s)	poison ivy	grain(s)	garden

Plants Bingo

cell(s)	carbon dioxide	algae	kingdom(s)	graft
biome(s)	photosynthesis	forest	producer(s)	chlorophyll
nectar	flower		pollen	stamen
bulb	moss(es)	botanist	vines	bush
grain(s)	coniferous	nut(s)	irrigation	tree

© Barbara M. Peller

Plants Bingo

bud	vegetables	botanist	photosynthesis	plants
perennial(s)	poison ivy	pollination	lichen	seeds
bush	roots		nectar	tree
petal(s)	coniferous	bulb	cell(s)	tropical rainforest(s)
chlorophyll	moss(es)	nut(s)	botany	pistil

Plants Bingo

bush	soil	deciduous	moss(es)	bulb
cell(s)	chlorophyll	vines	stamen	pollination
carbon dioxide	seeds		vegetables	botanist
fruit	grain(s)	garden	algae	pistil
herbs	biology	kingdom(s)	tropical rainforest(s)	roots

Plants Bingo

graft	Plantae	botany	chlorophyll	perennial(s)
fruit	bush	algae	nectar	biology
carbon dioxide	pistil		deciduous	photosynthesis
poison ivy	forest	vines	seeds	irrigation
vegetables	moss(es)	herbs	soil	biome(s)

Plants Bingo: Card No. 18

Plants Bingo

nectar	biome(s)	moss(es)	botanist	poison ivy
cell(s)	cactus (cacti)	photosynthesis	kingdom(s)	seeds
soil	stamen		annual	algae
irrigation	vegetables	herbs	coniferous	deciduous
pollen	plants	grain(s)	garden	nut(s)

Plants Bingo

poison ivy	soil	seeds	moss(es)	annual
bud	tree	lichen	herbs	perennial(s)
roots	pollination		petal(s)	deciduous
fruit	garden	agriculture	coniferous	vegetables
ferns	stem	plants	tropical rainforest(s)	forest

Plants Bingo

biome(s)	fruit	lichen	moss(es)	leaf
roots	deciduous	botany	botanist	flower
pistil	grain(s)		soil	algae
herbs	kingdom(s)	vegetables	cell(s)	garden
petal(s)	plants	nut(s)	bush	coniferous

Plants Bingo

pollen	irrigation	deciduous	producer(s)	bush
perennial(s)	seeds	graft	botanist	annual
botany	stamen		flower	pollination
vegetables	fruit	coniferous	forest	cactus (cacti)
plants	biology	soil	pistil	lichen

Plants Bingo

bud	soil	kingdom(s)	producer(s)	nut(s)
biome(s)	poison ivy	grain(s)	cell(s)	forest
irrigation	bush		agriculture	flower
pistil	stem	vegetables	biology	coniferous
leaf	garden	plants	herbs	deciduous

Plants Bingo

bud	poison ivy	cactus (cacti)	soil	botanist
nectar	nut(s)	lichen	perennial(s)	flower
pollination	photosynthesis		bush	pistil
leaf	agriculture	vegetables	biology	roots
ferns	petal(s)	plants	seeds	stem

Plants Bingo

petal(s)	lichen	soil	algae	deciduous
forest	leaf	cell(s)	bud	annual
roots	botanist		agriculture	vegetables
photosynthesis	fruit	stem	plants	seeds
nut(s)	cactus (cacti)	botany	chlorophyll	ferns

Plants Bingo

deciduous	soil	agriculture	perennial(s)	photosynthesis
irrigation	stamen	botanist	poison ivy	bud
leaf	herbs		seeds	petal(s)
bush	producer(s)	fruit	plants	vegetables
pollination	chlorophyll	algae	stem	ferns

Plants Bingo: Card No. 26

Plants Bingo

agriculture	botany	soil	poison ivy	tree
leaf	irrigation	cell(s)	vegetables	annual
vines	stem		plants	petal(s)
photosynthesis	biome(s)	ferns	lichen	forest
bush	seeds	deciduous	pollen	pollination

Plants Bingo

nectar	Plantae	tropical rainforest(s)	soil	botany
tree	deciduous	agriculture	herbs	seeds
stem	pistil		photosynthesis	perennial(s)
pollination	pollen	grain(s)	plants	vegetables
producer(s)	carbon dioxide	bush	ferns	leaf

Plants Bingo

deciduous	Plantae	photosynthesis	cell(s)	carbon dioxide
leaf	herbs	lichen	pollination	pollen
roots	soil		annual	agriculture
tree	fruit	graft	plants	vegetables
bud	botanist	ferns	biome(s)	stem

Plants Bingo

cactus (cacti)	soil	perennial(s)	carbon dioxide	vegetables
forest	photosynthesis	irrigation	seeds	annual
ferns	stem		pollination	lichen
leaf	biology	deciduous	plants	agriculture
fruit	kingdom(s)	biome(s)	Plantae	graft